The Secret Language
of Animals

The Secret Language
of Animals

JANINE M. BENYUS

Illustrations by Juan Carlos Barberis

FOREWORD BY
ALEXANDRA HOROWITZ

BLACK DOG
& LEVENTHAL
PUBLISHERS

THIS BOOK IS DEDICATED
TO THOSE WHO STRUGGLE
TO SAVE WILD LANDS

Published by
Black Dog & Leventhal Publishers, Inc.
151 West 19th Street New York, NY 10011

Distributed by
Workman Publishing Company
225 Varick Street New York, NY 10014

Manufactured in China
Cover and interior design by Elizabeth Driesbach

ISBN-13: 978-1-57912-968-2
h g f e d c b a

Library of Congress Cataloging-in-Publication Data available on file.

Contents

Foreword

I saw my first southern white rhinoceros in a zoo. I expect this mirrors many readers' experience, substituting *Sumatran tiger* or *polar bear* or *okapi* for my rhinoceros. For most people, that might be their last experience with the animal as well. Zoos are often the only place where most people encounter animals that are not pets, food, or considered urban or rural pests for their ubiquity (read: success). I have gone on to meet other rhinos, in their own habitats, but that first exposure stuck with me.

I was studying those zoo rhinos. In a graduate program of cognitive science, I found myself increasingly interested in the cognition of nonhuman animals, in what we could know about the abilities of species who cannot tell us what they know with words. I joined a series of research programs looking at animal behavior—a field called *ethology*—to equip myself for observing animals with an eye to making inferences about their cognition. One of these groups was watching rhinos; so, thus, did I.

It was a formative experience for me, not just in developing a greater rhino understanding, but in learning how to think about and to look at animals generally. The Wild Animal Park in Escondido, California, where I met these rhinos, is one of the zoos that has turned traditional zoo design a bit on its head: for much of this facility, the animals have large habitats (which they share with other appropriate species), and the humans are constrained to ride trains around the enclosures. As a researcher, I was allowed to sit on the banks of the hills outside the habitat, below the train tracks. But more important, I stayed on the hills for hours, whereas the trains paused only briefly—the trains' quick stops mimicking, really, the length of time a typical zoo visitor stands outside an animal's enclosure, gazing in. What I saw was that little of the animals' behavior could be witnessed from a passing train. The rhinos were visible from the train, but their true selves, their true behavior, was essentially invisible. For instance, as the trains loudly approached, the female rhinos, grazing together, would often stop what they *were* doing and orient themselves with their rumps together, heads outward, a defensive huddle against this unknown threat. They stayed frozen in this position until the train moved on, at which time they resumed their normal behavior. For the train visitors, I realized, the rhinos appeared to be "doing nothing;" of course, they weren't doing nothing. But to see what the animals were doing, one had to do two things: be informed, and loiter. That is, first, one had to know something about rhinos, for instance, that the species has poor vision; so as a defense mechanism, the animals stay together, each head alert to any possible predator to the group. Second, anyone hoping to learn something about rhinos has to stick around for more than a passing glance and really observe what the animals are doing, not just what one expects to see.

Observation, perhaps surprisingly, is not a simple skill. There is "looking," and there is *looking*—not just opening one's eyes and seeing, but seeing with fresh eyes, seeing without glibly assuming we know what's happening. Even outside zoos, animals are "behaving" all around us, but we mostly don't see what they are doing. To watch behavior carefully, and get the idea of what an animal is doing, it is critical that you have an idea who the animal is, where it is from, and what kinds of things it might want, need, and be able to do. An ethologist is not merely a watcher; she or he is a watcher with specific tools.

The Secret Language of Animals is one such tool that will allow the amateur ethologist to begin to truly observe animals. Benyus equips us with the language and the context to see things we otherwise would miss. We learn both how to look at an animal's behavior and body to determine where the animal lives and how to do the reverse: to use the animal's environment to make sense of its behavior and the design of its body. We learn the meaning of the gorilla chuckle and the lion rump-swivel, the reason for a giraffe neck-slap, and the significance of a panda honk. Armed with one of her case studies of animals before your next zoo outing, your visit will be transformed.

It was a privilege for me to watch those great animals, the rhinoceroses, as it is a privilege to be able to gaze upon the lives of all animals, zoo'd, wild, or domestic. What I have seen has changed me. The "secret language" of animals is decodable for those who know how to look. The pages that follow hold your code. Go observe!

—Dr. Alexandra Horowitz,
author of *Inside of a Dog* and *On Looking*

Acknowledgments

To gain perspective on a subject this vast, a writer must stand on the shoulders of giants. Hoisting me aloft were the thousands of researchers who braved white-outs, monsoons, bone-chilling nights, and scorching noons to bring home news of animal behavior. I am indebted to them, as well as to the zoo professionals who shared with me their enthusiasm and fierce devotion to wildlife conservation.

The technical reviewers were Dr. George Archibald, Director, International Crane Foundation, Wisconsin; Dr. Cheryl Asa, Reproductive Biologist, St. Louis Zoo; Dr. John Behler, Curator of Herpetology, Bronx Zoo; Dr. Daryl Boness, Research Zoologist, National Zoo; Dr. Donald Bruning, Curator of Ornithology, Bronx Zoo; Teresa DeLorenzo, Research Affiliate, Metro Washington Park Zoo, Portland; Dr. Sue Ellis-Joseph, Conservation Education Specialist, Minnesota Zoo; Dr. Susan Evarts, Post-Doctoral Associate, Bell Museum of Natural History, University of Minnesota; Dr. Martha Hiatt-Saif, Senior Trainer, New York Aquarium; Dr. Brian Joseph, Veterinarian, Minnesota Zoo; Dr. Jeffrey Lang, Professor, University of North Dakota; Dr. Frank McKinney, Curator of Ethology, Bell Museum of Natural History, University of Minnesota; Dr. Jill Mellen, Conservation Research Coordinator, Metro Washington Park Zoo, Portland; Dr. Jackie Ogden, Research Fellow, Center for Reproduction of Endangered Species, San Diego Zoo; Lisa Rappaport, Behavioral Researcher, Riverbank Zoo, Georgia; Bruce Read, Curator of Mammals, St. Louis Zoo; Dr. Michael Robinson, Director, National Zoo; Conrad Schmitt, Associate Curator, Cheyenne Mountain Zoo; Peter Shannon, Curator of Birds, Audubon Park, Louisiana; Dr. Steve Sherrod, Director, Sutton Avian Research Center, Oklahoma; Dr. Beth Stephens, Research Biologist, Zoo Atlanta; Dr. Lisa Stevens, Mammal Collection Manager, National Zoo; and Scott Swengel, Assistant Curator of Birds, International Crane Foundation, Wisconsin. Their guidance was invaluable, and I thank them.

The Secret Language and Remarkable Behavior of Animals was lovingly birthed by talented midwives. I was fortunate to work with visionary agent

Jeanne Hanson, editor extraordinaire Nancy Miller, the illustrious Juan Carlos Barberis, the ever-calm production maestro John Fuller. The fact that these people cared about the book made the whole process a pleasure.

Although writers are not a particularly social species, I have been blessed with a warm circle of friends and family. My thanks to Cynthia Robinson for lending a generous ear during the early stages. For typing gigabytes of notes, editing with a fine eye, and buoying me with her company, I am deeply thankful for Mary Ann Hatton. Laura Merrill, who was as constant as the northern star, gave me the immeasurable gift of hope and laughter, for which I cannot begin to thank her enough. Laura also pored over the galleys, as did Daniel Benyus and La Rue Moorhouse. Even Eight Ball, the enigmatic tomcat who crowded me at the keyboard, played a part. Any errors or pearls of wisdom found on these pages were probably typed in by his paws.

Finally, I want to thank the moonlit mountains of Montana. Living here, I've learned that there really is no place like home and that every last, wild acre is worth protecting.

The Secret Language
of Animals

A koala sleeps soundly
in its eucalyptus home.

There's No Place Like Home

It was the wee hours of the morning when Neil Armstrong took his first small step on the moon. My parents and sister, after gallantly waiting up for hours, had yawned their last yawn and fallen asleep. I was only 11, but I was wide awake in front of the television, anxious not so much for Armstrong to set foot on the moon as for him to turn around and tell us what the earth looked like from there.

I know I wasn't the only one. Twenty-three years later, I've noticed that we don't hang pictures of the lunar landscape on our office walls. It's not the Sea of Tranquillity that distracts us from our work, but that homeward-bound shot—that misty, turquoise globe sailing all by itself in a sea of black. Like any good symbol, the earth's portrait speaks straight to the heart.

Look, it says, we're all in one boat here, and it's the only one we've got.

Knowing that hasn't stopped us. In a few reckless decades, we've punched a hole in the ozone above Antarctica and sprung thousands of other environmental leaks that are just as hard to fix. Reaching the moon didn't put our problems any farther behind us. All it made us realize was how unusual the earth is, how it alone contains the right levels of light, temperature, gases, and water to keep us alive. This is where we evolved, and every cell in our body is designed to cope with and take advantage of this place. Even if we could find a similar planet to beam ourselves up to, wouldn't we miss the plants and animals and even the insects we grew up with?

After all, we're not the only ones who evolved in this petri dish of soil, sun, air, and water. Millions of species live here, and all of them, from amoebas to zebras, were shaped by the peculiar demands of their home habitat. The earth turned out to be a wildly diverse place. Rain puddles, jungles, ice caps, and searing deserts were mothers to very different sorts of biological inventions.

The wonderful thing about a zoo is that you can see an entire globeful of nature's inventiveness in a few accessible acres. You can see, smell, hear, and be splashed by the earth's best and brightest, the products of a cornucopia of habitats. In the new zoos, exhibit designers are working hard to recreate these habitats, both for our education and for the animal's well-being. They've found that the best way to call up the full range of natural behaviors is to put animals in the environments they evolved in. What a tall order! An arctic icescape in San Diego, a coral reef in Minnesota, a steamy tropical jungle in the Bronx. These exhibits are amazing, to be sure, and the animals seem more at home here, which of course delights the designers and zookeepers. But, like space travel, it also humbles them.

Planning for life on another planet is a lot like exhibit design; you quickly learn how hard it is to re-create a habitat. You can choose the right plants and animals, but how do you stage the interplay of soil and water, predators and prey, allies and enemies? How do you re-create the perfect, perplexing balance of it all? In the end, it's not how habitats look but what they do that is magical. Wild, healthy habitats do more than support life; they challenge it, inspiring new life forms and fine-tuning the ones that already exist. No matter how sophisticated our exhibits become, we can't touch that.

Maybe that's why so many people in the zoo world are committed to saving the genuine article. They know that an animal is part and parcel of the forest, grassland, desert, or ocean that it evolved in. Without its native habitat, even the fastest, strongest, most clever organism is a sorry fish out of water. By the same token, if we stop and think what will happen if we finish off the rain forests, eat up more ozone, and keep on reproducing like there's no tomorrow, something dawns on us. We realize that we're just as vulnerable as the disappearing sea turtles, spotted owls, elephants, and manatees; we're as bound to this irreplaceable place as they are. If we suddenly found ourselves out of home waters, we'd be gasping for breath too.

Ask an astronaut.

When I look back, I think maybe the real Giant Leap for humankind was for a few people to travel far enough away from earth to feel homesick for the first time. The planet they flew back to is not as serene as it looked from a distance.

It is warping under the G-force of too many people, and among those feeling the effects most keenly are the incredible animals you will meet in these pages.

As you read their stories, I hope you'll be entertained and enamored and feel like staying up late to learn about them. But even more than that, I hope you'll feel a little homesick for the paradise that these animals evolved in. I'd like to think it's not too late to patch our holes, bail like crazy, and make this turquoise vessel seaworthy again. In the meantime, we need to honor and protect all the species that still sail with us and dedicate our next Giant Leap to them.

Housed in habitats instead of cages, today's zoo animals are displaying more natural behaviors than ever before.

What's New with Zoos?

If you haven't been to a zoo in several years, you're in for a wild surprise. Zoos have weathered a tough soul searching over the last few decades, and the good ones have re-created themselves from the inside out. They've sprung the cages and turned the animals loose in startling simulations of their home habitats, some so lifelike that you'll swear you're being stalked by that leopard in the leaves or that wolverine on the hill. The authenticity seems to agree with the animals as well. They get to vine-swing in tropical forests, dive in living coral reefs, dabble in creeks, and burrow to their heart's content in prairies. In a wonderful turnabout, it's the zoo visitors who are now hemmed in by railings, not the residents.

What a welcome change! Instead of feeling your heart sink at the sight of a despondent gorilla, you'll feel it race as a barrel-chested silverback explodes into view, then disappears in the greenery. You may have to search for him there, but that's a good sign; it means gorillas and other animals are blending with the landscape in a way they never could when marooned on tile floors and manicured lawns. For the first time, zoo animals have the space and privacy to prowl, howl, court, build nests, and defend their territories. Besides being more at home, the animals are also in better company. No longer the lone representative of their species, they now romp in herds and pods, troops and bevies. Some have even decided to put down roots, and if you search closely, you'll see kits, calves, joeys, and cubs, some of which are the first of their kind born in captivity.

Although these exhibits make zoos all the more entertaining, their real agenda is to educate. By immersing us in the animal's world, they show us how the animal evolved in tandem with its habitat and how, from crest to claw, it is adapted to live where it lives. The exhibits also bring out natural behaviors in the animals, prompting them to act more like themselves than they ever have in captivity. In fact, scientists who once scorned zoos will often bring binoculars and clipboards to study animal behavior close up. The findings are helping zoos fulfill what has become their foremost mission: to successfully breed the endangered species that have landed in their lifeboat.

Just how did zoos move from their days of bars-and-shackles to the look and outlook of today? To answer that, we have to rewind a few million years, to the very beginnings of our ancient relationship with animals.

Our Love Affair With Animals

It was no more than an evolutionary eyeblink ago that our apelike ancestors were crawling on their hands and knees in pursuit of big, dangerous, delicious animals. For five million years (99% of our time on earth), our survival hinged on finding animals that we could eat. Stalking an animal meant knowing everything about its habits: where it slept, where it drank its daily water, how fast it would run once it caught wind of us. By necessity, we became astute observers of our fellow creatures and learned a respect that comes with intense study of

Thousands of generations and evolutionary steps later, humans are still fascinated by wild animals.

a subject. Although we've stepped farther and farther from the wild in our last 5,000 years of agriculture and industry, an ancient awe still stirs deep within us. It's that to-the-bone shiver we feel when we see a wolf trot lightly from the woods or hear the thin song of a rising whale.

What is this uncanny ability animals have to amaze, delight, and at times frighten or repulse us? Part attraction, part fear, and part admiration still make us curious about wild animals even though we no longer need them to fill our stomachs. In North America alone, 134 million people pour through the turnstiles of accredited zoos and aquariums each year. That's more people than go to all the professional sporting events combined! Worldwide, at least 700 million visitors—10% of the entire world population—attended the 1,500 zoos and aquariums united in the World Association of Zoos and Aquariums—that's more than all the people in the United States, Canada, France, and the United Kingdom. And it's not just school groups either. For every wide-eyed child, there are three astonished adults rushing to the rail to see the dolphins leap.

These days, animals in zoos represent more than just a shadow of our ancestral past. They are the last ambassadors of a world that is rapidly becoming less and less wild. The red-alert sirens are screaming, and, for the first time in history, we are admitting to our destructiveness and grappling to right the wrongs. Zoos that were once primarily amusement parks are now on the front lines of that fight, working to brighten the future for rare animals in their keeping. To track this sea change in the zoo world, we have to go back thousands of years, to the creation of the very first zoos.

Outgrowing the Old Zoos

Zoos were originally menageries kept by royalty as a show of power and wealth. After all, if you could arrange to ship a giraffe 1,500 miles down the Nile (as Queen Hatshepsut of Egypt did 3,500 years ago), you could probably get anything you wanted. Thousands of years later, when zoos finally opened their doors to the public, they became sources of pride for local communities. The goal was to have one example of as many different species as possible, like a stamp collection of animals. Various zoos guarded their collections jealously, and if one zoo knew the secret of, say, keeping a rare turtle alive, it wouldn't dare tell another zoo. With each collecting trip to Africa or Asia, the zoo's acquisitions grew. It didn't matter that the animals were confined to cramped, barren cages or that they often died of disease or stress; it was easy enough to get replacements from the wild. Although the majority of the public stayed silent about the bad conditions at zoos, not everyone was satisfied.

A small revolution began in the early nineteen hundreds when Carl Hagenbeck of Germany did away with the cages and put his animals on wide green lawns surrounded by hidden moats. Visitors applauded the change, but it took some time for the exhibits to catch on in other zoos. One of the problems was that zookeepers found it hard to control diseases that spread in the soil and grass of outdoor exhibits. A cement box was easier to hose down, and advocates claimed it was better in the long run for the animal's health, if not for its spirits. After World War II, medical technology improved and so did animal health care. Many zoos began to feature Hagenbeck-like exhibits, at least for their hoofed animals and cats of the African plains. Still, the majority of zoos put animals behind bars and replaced their losses with captures from the wild. These zoos enabled people to see animals, but instead of engendering respect, the pacing jaguars and psychotic polar bears often provoked pity and a feeling of guilt among visitors.

In the 1960s, this collective guilt became sharper, focused through the lenses of writers, activists, and opinion-makers. People were becoming aware that we share this planet with other creatures that have every right to be here. As our consciousness of animal rights matured, true animal champions suddenly had a dilemma. They wanted to see wild animals and reconnect with their roots, but the deplorable conditions in many zoos kept them away. Some took issue with the restricted quarters and with the practice of capturing animals from the wild. Others protested the heavy emphasis on vaudeville-style entertainment such as chimp tea parties, penguin parades, and walrus ballets. These rumblings of discontent came not only from outside, but also from within the zoo staffs themselves. It became clear that zoos were losing credibility and would soon be forced to evolve or to shut their gates for good.

Big Shoes to Fill

Meanwhile, as the debate about captive animals raged, sobering news about their wild counterparts started to trickle in. Scientists reported that many of the species thriving in zoos were just barely hanging on in the wild. For some species, they predicted that the numbers in captivity would soon be greater than those in the wild. Overnight, it seemed, zookeepers were caring for some of the most precious cargo on earth. Realizing this, zoo directors and conservationists from around the world started to draft a whole new role for zoos.

The days of collecting wild specimens were over. Zoos began to experiment with breeding, and by the mid-1980s, more than 90% of zoo mammals were born in captivity. Although this meant that zoos could resupply themselves rather than deplete wild populations, it was not enough. As populations on the outside teetered, zoos began to realize that they might someday host the only examples of a species left on the planet. When that day came, the responsibility, not only for individuals, but for an entire species—its physical and genetic health and its possible reintroduction to the wild—would rest squarely on the zoo world's shoulders.

That day has come. Extinctions are occurring at a rate unprecedented in the planet's history, rising from a loss of one species every 5 years in 1850 to the current rate, according to the U.N. Environment Program, of 150 to 200 species of plants, mammals, birds, and insects every 24 hours. You may notice the proliferation of Vanishing Species signs in front of the animal exhibits at your zoo. The prognosis is not good for many of these faltering species; experts predict that as many as 30 to 50 percent of all species could be extinct by mid-century.

Preventing extinction isn't just a matter of protecting animals from poachers. The real problem is that wild animals are running out of space, and, where they do have space, their habitats are being degraded. It's hard to find a place on earth that hasn't felt the grip of human greed and its alibi, the so-called license to "subdue and dominate." Even the woolliest of wilderness are now surrounded by a lasso of development that cinches tighter each day. Tropical deforestation (at 150 acres a minute) combined with pollution (warming the world by approximately 1.53 degrees Fahrenheit since 1880) and the prolific human population machine (producing more than three hundred and fifty thousand new people a day) threaten to push many species over the edge. In fact, conservative estimates say that 25% of all species will be in trouble within the next few years. Faced with statistics like that, zoos have become modern-day arks, called as Noah was to protect the future of entire evolutionary lines.

Protecting and breeding these animals, as important as that is, is just the first part of the solution. The ark is only as good as the promise that someday the flood will subside and we can release the passengers to suitable habitats. Unfortunately, suitable habitats are getting harder and harder to find. If the ark is to be anything more than a pipe dream, we must devote ourselves whole-heartedly to the other part of the solution: we must stop the flood of habitat destruction at its source. The fact is, after all is said and done, a species of wild animal doesn't really belong in a zoo. It belongs in the wild, living by its wits and evolving in the face of natural challenges.

Zoos are in a unique position to tell this story to millions of people who are smiling, enjoying themselves, and primed to learn and listen. Zoos can tune us in and turn us on. They can make us angry about the assault on wild habitats and show us how to harness our energy in votes, dollars, lifestyle changes, and volunteered talent. In the meantime, while we struggle to salvage what's left of the wild, the new zoos can act as safety nets, centers for research, and places where we can enjoy the exquisite pleasure of spending time in the company of animals.

Zoos as Safety Nets

Biologists agree that the only way to really save animals is to save their habitats. Although bits and pieces of wild land are being protected throughout the world, the current trend is a net loss of habitats. Even the 1% of the world that is set aside in national parks and refuges is not free of encroachment by poachers and people who are desperate for food and fiber. As writer Colin Tudge reminds us, hungry people have no time for rare animals.

The human population passed a landmark of seven billion in 2011. Demographic experts say that if we manage to avoid epidemics, wars, and ecological collapse, there will be 10 billion of us on earth by 2100. At that point, they predict, our population will level off for five centuries and then begin to decline, relieving the pressure on what is left of the land. Here's where zoo people prove to be true optimists. They hope to maintain a reservoir of animals and a bank of frozen sperm and eggs so that habitats, if they recover, can be repopulated. On the fragmented habitat "islands" where animals have survived, sperm and eggs from zoo animals could breathe fresh genetic air into stale populations.

There are plenty of pessimists who have no hope that any of this will be possible. They point to high mortality rates in previous attempts to reintroduce species to the wild. After living in captivity, they say, predators will forget how to hunt, and prey species will lose their wariness. Imprinting on humans may cause animals to have no fear of human hunters and perhaps leave them unable to have healthy sexual relationships with members of their own species. Captive breeders counter all these claims, pointing to examples of feral dogs that have returned to the wild after being house pets for several generations. To ease this reentry for zoo animals, breeders utilize reintroduction training programs and try to raise young without imprinting them. Much research and development has gone into improving these programs over the past two decades. Although they admit that many captive-bred animals will die when released into the wild, they feel that even a small survival rate is better than none.

Though captive breeding is a number one priority at many zoos, it is not always obvious to the casual visitor. Much of the work is done behind the scenes at off-site facilities like the Smithsonian Conservation Biology Institute's Front Royal Campus in Virginia. Here animals are raised outdoors in proper social groups and bred according to a sophisticated genetic plan that involves not only the Center's animals, but animals at zoos throughout the world. You see, unlike Noah, we can't simply herd the animals two by two onto the ark and pray for sun. If we hope to release these animals into the wild someday, we have to keep them genetically sharp enough to be able to react to changes in their environment. A species is at its best when enough individuals are breeding to keep the gene pool diverse. Having a diverse gene pool is like having a large toolbox filled with tools that you don't use daily but that someday you might need. If the population of breeding animals is too small or genetically limited, it leaves fewer tools to solve problems with, should disaster strike. Besides having less genetic ingenuity, small populations may start inbreeding, which can perpetuate rare genetic weaknesses.

Thanks to captive breeding efforts, rare creatures such as mountain tapirs are giving birth to a new generation.

To avoid these pitfalls, zoo managers have cooperated in a worldwide strategy called the Species Survival Plan (SSP) since 1981. Today there are more than 500 SSP programs in place to help manage endangered species populations. For the purposes of breeding, all captive rhinoceroses, for instance, no matter where they reside, are considered part of the same population. Stud records, showing who was born to whom, are stored in a database called ISIS (International Species Information System), housed in Minnesota. When a female is ready to go into heat, ISIS is used to find a male with "fresh" genes, even if he must be flown from Moscow to Chicago for the occasion. The goal of this family planning is to enlarge the breeding population in captivity and then subdivide the population among zoos (a precaution against epidemics). If the matchmakers at ISIS have their way, the genetic structure of zoo populations will resemble that of wild populations, so that one day, we'll have a relatively intact species, instead of just individuals, to release into wild areas.

Knowing what we must do is not the same as being able to do it, however. Studies have shown that a full fifth of the world's vertibrates are currently at risk of extinction. The number of at-risk invertebrates, which are even more critical to ecosystem health, could soar into the millions. Keeping viable populations of even a portion of these species would be a formidable task given the small size and budget of most zoos. Consider, for instance, that all the zoos in the world could fit into the borough of Brooklyn, New York, a mere 20,000 acres. Space considerations aside, yawning gaps remain in our knowledge of rare animals. We probably know as much about breeding these soon-to-be-extinct creatures, for example, as people knew about raising livestock 4,000 years ago. Even the technology of sexing animals (determining their gender) to allow us to find out whether we have males or females in our bird and reptile collections has only been available since the 1980's.

A current practice is the establishment and maintenance of "frozen zoos." These are facilities where sperm, embryos, and other genetic material of animals can be preserved at very low temperatures. Even before filling the test tubes, however, we have to decipher the basic breeding cycles of rare creatures, a job that calls for coordinated record keeping among the world's zoos. In years past, record keeping was often a hit-or-miss operation, and sharing information was unheard of in the competitive atmosphere of zoos. Today, using ISIS, we have been able to assemble family trees for nearly 10,000 species and have hopes of adding new ones as our knowledge and store of records increase. In the meantime, there's a lot to learn about how animals reproduce, and not much time left for many species. In the shadow of the ticking clock, the new zoos have inherited yet another important mission: to be centers of animal behavior research.

What Zoo Animals Can Teach Us

As the zoo world repairs its reputation, a wonderful resource is being taken out of its shroud. For years, scientists studied the remains of dead animals at museums, scrutinized lab animals, but simply overlooked the opportunities at their local zoos. Here they had a chance to study live animals at close range— being born, growing, learning, resolving conflicts, building a home, winning a mate, parenting, and aging. Those who did take advantage of zoo studies laid legendary groundwork in the field of animal behavior. It was in a zoo setting, for instance, that the facial expressions of wolves were first studied in detail. Given the skittish nature of wolves, this subtle "language" would have been nearly impossible to decode in the wild. The panda was equally difficult to study in the wild. Its solitary habits and remote, forested haunts kept researchers from learning about reproduction—until the first panda cubs were born in zoos.

With the advent of lifelike immersion exhibits over the past several years, animals are rounding out their repertoire of natural behaviors, making zoo study all the more revealing. Behaviorists get an eyeful, since zoo animals generally have more time to devote to social interactions than wild animals do. Courtship, mating, and parenting are especially fertile subjects, given that animals are not quite as secretive as they would be if predators were afoot. Close-up looks at development and anatomy are also naturals for zoo research, along with studies on nutrition and on animals' reactions to captivity.

There are, of course, limitations to what we can study at zoos. Because zoo animals are fed, inoculated, and shielded from harm, some of the most important ecological facts of life are missing from the equation. We can't, for instance, study predator-prey relationships, migration, or animals' reactions to seasonal shortages of food. This book describes the behaviors we can see in zoos: the everyday routines of feeding, body care, and movement, plus social behaviors such as bonding, aggression, courtship, and parenting. Naturally, some animals perform these behaviors differently in their exhibit than they would in the wild, depending on how sensitive they are to human presence. Even when you factor in the human influence, however, zoo-based studies can still provide valuable baseline data for field studies. These results can also be used to manage semiwild populations in the outdoor megazoos called national parks.

Far from their home turf, zoos have spearheaded and funded some of the most famous studies of wild populations. In turn, data from these studies, such as Jane Goodall's work on chimpanzees, are often used to design zoo exhibits, closing the circle and uniting field researcher and zoo professional in a spirit of cooperation. This spirit is absolutely essential if we are to scale the walls that face us. As part of this effort, the last and most pressing mission of zoos is to recruit the public's help in stemming the tide of habitat destruction.

Recruiting Everyone's Help

Besides being places where animals can breed and scientists can study, zoos are the only places where most people can watch, hear, smell, and meet rare animals in living color. All the television shows, museum dioramas, and encyclopedias in the world can't match the chemistry that occurs when animals and people look into each other's eyes. This chemistry works its own magic, touching people's hearts in a way that lasts. For many children, a trip to the zoo is their first real encounter with the animals they have read about, sung about, and drawn

Meeting an animal face-to-face can awaken a lifelong love of nature.

since their earliest years. The fact that animals figure so prominently in children's fables as well as in mythology, art, and language says volumes about our connection to wild creatures. At the zoo, myth becomes reality, and the connection is reaffirmed in a new way.

Spending time with an animal in a naturalistic exhibit puts us in its world, kindling an affection and, eventually, a concern for the animal's well-being. Like our ancestors, we want to know more about the animals we watch: where they live, how they live, and whether they will survive the crushing assaults on their wild habitats. Happily, learning happens naturally at a zoo; the animals pique our interest, and the exhibits, if they are well done, quench our thirst for information.

Zoo people know that the crucial step in the making of a conservationist, after awareness and education, is a commitment to change the status quo. It is at the precise moment when people are falling in love with animals that they are best able to hear the cry for help and respond in a personal way. As the entrance sign at the Bronx Zoo says, "In the end, we will conserve only what we love. We will love only what we understand. And we will understand only what we are taught."

The Future of Zoos

After years of writing exposés about the sad and shocking decline of zoos, the media is now trumpeting the zoo renaissance. Indeed, zoos around the world have spent hundreds of millions of dollars the last several years to improve living conditions for animals and to educate the new generation of zoogoers. People who were once uneasy at zoos are breathing a sigh of relief, and zoo attendance continues to rise.

What will be even more fantastic—and in some senses, more controversial—are the zoos of tomorrow, now on the drawing boards of a few visionary individuals. Biologists are looking ahead to the use of cloning to reproduce endangered species, as well as bioengineering more docile animals for fenceless exhibits. Exhibits such as the "Arctic Ring of Life" at the Detroit Zoo push the boundaries of indoor and outdoor exhibits by recreating the environment of the tundra, allowing polar bears, arctic foxes, and seals to inhabit a shared space.

Making the ordinary extraordinary will be one of the greatest challenges of the future zoo, legendary National Zoo director Michael Robinson once said. Long-running, award-winning exhibits such as one devoted to the beaver at the Minnesota Zoo demonstrate that zoogoers want to learn more about the commonplace creatures of their regional habitats. Though the beaver is not itself

A day at a habitat-oriented zoo is a journey through unforgettable neighborhoods like this bustling prairie-dog town.

endangered, some of the species that use its habitat are. Only when people can see that water beetles, ferns, and spotted frogs are as fascinating as Siberian tigers can we begin to save not just a handful of species, but the entire biome, the wellspring of life.

Wandering through bioclimatic zones of plants and animals will be a far cry from shuffling past row upon row of caged specimens as we did in the past. After a day of absorbing wild creatures in their lush natural contexts, we'll feel as if we've traveled from mountaintop to ocean floor and back again. That night, as we drift off to sleep in our own home habitats, it will be the connections between living things that we remember and the fact that our place in the puzzle is not on top, or apart, but somewhere in the middle.

This book is designed to be your companion not just in today's best zoos, but also in the bioparks of tomorrow. In these pages, you'll find animals "exhibited" in their bioclimatic zones, and you'll read about what they do in the wild as well as in captivity. Zoos are changing, and I'm hopeful that even if you can't see all these wild behaviors today, you'll at least see glimpses of them. More importantly, you'll know what to look for in the days ahead when, as exhibits get wilder and wilder, what you see at the zoo will too.

INSECT-EATERS. Chameleons use gripping feet, a long sticky tongue, and phenomenal balance to garner a bit of protein.

How Animals Behave: A Primer

Why Animals Behave the Way They Do

WHAT IS BEHAVIOR?

Behavior is a survival maneuver. It's everything and anything an animal does to keep itself alive today and its genes alive tomorrow. Building a home, courting a mate, finding food, stalking prey, threatening an enemy, and soaking up the sun's warmth are common survival moves, but, as you will see in this book, each species performs them with its own flair. These differences distinguish a mongoose from a marmot and tell us volumes about where an animal lives and what it has to face each day.

When we visit animals in zoos, it's sometimes hard to remember that they evolved in wild habitats where they were not fed by zookeepers, assured of shelter, fenced from enemies, or checked regularly by the zoo vet. Yet it was the harsh reality of limited food, weather, predators, and diseases that shaped their bodies and behaviors in the first place. Though the struggles in a desert are different from those in a tropical jungle, the principle behind them is exactly the same. So before we talk about what makes animals different from one another, let's celebrate what unites them.

A CRACK AT IMMORTALITY

If there's one thing that all of us—from slime molds to slide guitar players—have in common, it's DNA. Units of DNA called genes contain blueprints that could be used to recreate our bodies from scratch if need be. In fact, when an embryo is starting to grow, it follows the instructions on those genes to a tee. In asexual reproduction, or cloning, the parent simply creates an identical copy of its genetic blueprint; barring a rare mutation, the young is a carbon copy with no new features. In sexual reproduction, however, two individuals conspire to create a new blueprint made up of one-half of the mother's genes and one-half of the father's genes. This conspiracy allows new genetic combinations, new instructions, to emerge, paving the way for innovative new features.

The animals we share the earth with today are here because the combinations that cropped up were winners. They gave rise to new physical and behavioral features that gave animals an edge over other contestants in their habitat. This edge enabled the recipients to live longer and have more offspring than their competitors, making them, in the parlance of natural selection, the fittest of the fit. Because their winning designs were blueprinted on genes, animals could pass these advantages on to their offspring, and they in turn passed them on to the next generation.

Because genes live on in the offspring when a parent perishes, genes have a crack at immortality, or, at least, a longer tenure than the bodies that carry them. This gene immortality idea gives us one way of understanding why animals behave the way they do. Richard Dawkins, in a remarkable book called *The Selfish Gene*, says animals do what they do not because the animals themselves want to survive, but because the genes within the animals "want" to be replicated and thus increase their frequency in the larger gene pool. Our bodies are survival machines—carrying, protecting, and promoting our genes. As Samuel Butler once said, a hen is the egg's way of producing more eggs. And we, says Dawkins, are the hens for our genes.

This concept makes a lot more sense when you imagine what would happen to genes that built a survival machine without the behavioral and physical tools for survival. The faulty machine would die out, perhaps without leaving offspring. Before too long, those genes would disappear from the gene pool, while at the same time, the genes that were building better and better carriers would be getting more numerous.

It's an interesting idea, but how can it help us understand what zoo animals are doing? Simple. Whenever you see a behavior, think to yourself, is this behavior adaptive or nonadaptive? If it's adaptive, it must somehow help the animal's genes increase their frequency in the gene pool. Does the behavior make the animals more attractive to the opposite sex? Does it help them deal with food shortages in winter? Does it allow them to sneak by dominant animals without a scratch? Does it enable them to eat and avoid being eaten? If the behavior jeopardizes the animal's chances of passing on its genes, it may be what's called nonadaptive. Eventually, if a nonadaptive behavior is detrimental enough, the animals that carry it will do poorly, and the genes that control it will be edited out of the gene pool.

On the surface, some perfectly adaptive behaviors look as if they would be detrimental for the animal's genes. Let's take an example. Say you are watching the chimpanzee exhibit and you notice that a mother gives up her baby for a while to another chimp, who baby-sits and even nurses the infant. This helpfulness seems to contradict everything we've said about the selfish gene. What do these nannies have to gain by helping young that are not their own? The answer goes back to the common denominator of life, the genes.

Because chimps travel in troops of closely related individuals, the baby-sitter is likely to have some genes in common with the mother. If she is a sister or a daughter, 50% of their genes are the same. Since some of the helper's genes are bound to be present in the infant, the helper does have a valid genetic stake in its survival. She is also getting good practice for the day when she will raise her own infant, perhaps more successfully because of her early experiences. Once this helping behavior proves beneficial, the gene that controls or influences it will continue to be passed on, making the chimp an even better survival machine.

By knowing the genetic realities that drive animals, we can avoid slipping into anthropomorphism, or attributing human traits to animals. Think, for instance, how easy it would be for us to say, "Look, that female chimp is doing that new mother a favor. Chimps must have big hearts." This would be denying the fact that wild animals, unlike most modern-day humans, have to fight for their survival. The genes that support survival will live on, whereas the genes that detract from it will eventually be edited out. And, as genes go, so goes animal behavior. In the end, genes pull the strings, dictating how and when an animal behaves the way it does.

Some people may say that thinking about animal behavior in terms of genes somehow takes away its mystery, but I think it makes the riddle more enticing. As Marion Stamp Dawkins, author of *Unravelling Animal Behavior*, says: "The fight for a place in the genepool has given rise to some of the most beautiful and intricate phenomena on this earth. Animals…have developed the power to swim and to fly, to care for their young, to stalk their prey, to play, to sing, and to be curious about the world around them. To know all this comes from such simple beginnings can enhance and deepen the winder, not diminish it."

THE ROLE OF HABITAT

Although the selfish gene idea helps us explain the common motives behind behavior, we need something else to illuminate all the differences that make animal watching so absorbing. If the ultimate goal is the same throughout the world, why is there such marvelous variety in the way animals go about surviving? The simplest answer is this: though the game is the same, the playing fields—or habitats—are different throughout the world.

A habitat is the place where an animal finds what it needs to survive. More than a geographical place, a habitat is all the opportunities and challenges an animal faces, including competition, climate, food availability, predators, and a host of other conditions that can't be described on a map. Through the wonders of natural selection, each survival machine is custom crafted to excel in its particular constellation of living conditions. Its body and behavior echo the habitat it evolved in.

Figuring out where an animal evolved takes a bit of detective work. We do it all the time when we try to guess which part of the world someone is from. We look at a person's mannerisms, dress, and accent to figure out where he or she grew up. In the same way, an animal's mannerisms are a clue to its homeland. An animal that climbs up the exhibit when frightened, for instance, probably lived its natural

ADAPTATION TO HABITAT.
An animal's body echoes its habitat.
This tree-dwelling barred leaf frog has
suction-cup toes and light-gathering
eyes to help it maneuver in its
leafy, shaded habitat.